The Solar System

Robin Birch

Neptune

CHELSEA
CLUBHOUSE

An Imprint of Chelsea House Publishers
A Haights Cross Communications ☙ Company

Philadelphia

This edition first published in 2004 in the United States of America by Chelsea Clubhouse, a division of Chelsea House Publishers and a subsidiary of Haights Cross Communications.

Chelsea House Publishers
1974 Sproul Road, Suite 400
Broomall, PA 19008-0914

The Chelsea House world wide web address is www.chelseahouse.com

Library of Congress Cataloging-in-Publication Data Applied for.
ISBN 0-7910-7930-9

First published in 2004 by
MACMILLAN EDUCATION AUSTRALIA PTY LTD
627 Chapel Street, South Yarra, Australia, 3141

Associated companies and representatives throughout the world.

Edited by Anna Fern
Text and cover design by Cristina Neri, Canary Design
Illustrations by Melissa Webb
Photo research by Legend Images

Printed in China

Acknowledgements

The author and publisher are grateful to the following for permission to reproduce copyright material:

Cover photograph of Neptune courtesy of Photodisc.

Art Archive, pp. 5 (bottom), 22 (right), 25 (right); Australian Picture Library/Corbis, p. 24; Digital Vision, pp. 13, 27; Calvin J. Hamilton, pp. 7, 11; Walter Myers/ www.arcadiastreet.com, p. 10; NASA/JPL, pp. 5 (top), 12, 20, 22 (left); NASA/NSSDC, pp. 19, 24; Photodisc, pp. 4 (right), 17, 28; Photolibrary.com/SPL, pp. 6, 14, 15 (all), 18, 21, 23, 25 (left), 26, 29.

Background and border images, including view of Neptune, courtesy of Photodisc.

Please note
At the time of printing, the Internet addresses appearing in this book were correct. Owing to the dynamic nature of the Internet, however, we cannot guarantee that all these addresses will remain correct.

Contents

Glossary words

When you see a word printed in bold, **like this**, you can look up its meaning in the glossary on page 31.

Discovering Neptune

People in **ancient** times did not know about the **planet** Neptune, which was discovered in 1846.

Before Neptune was discovered, **astronomers** were finding that the planet Uranus did not go where they thought it should, as it moved among the **stars**. Some of them decided an unknown planet must be affecting Uranus's **orbit**.

John Adams and Urbain Leverrier worked out where the unknown planet should be in the sky. When Johann Galle searched in these places with a **telescope**, he found Neptune.

▲ This is the symbol for Neptune.

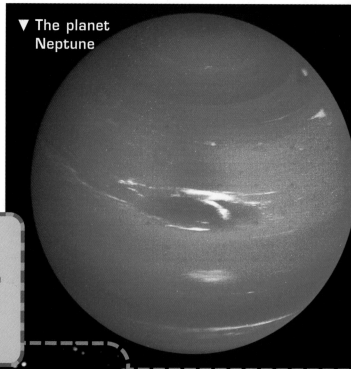

▼ The planet Neptune

John Adams was an English astronomer, born in 1819.

Urbain Leverrier was a French mathematician, born in 1811.

Johann Galle was a German astronomer, born in 1812.

The word "planet" means "wanderer." Stars always make the same pattern in the sky. Planets slowly change their location in the sky, compared to the stars around them. This is why planets were called "wanderers."

▶ Neptune with its largest moon, Triton

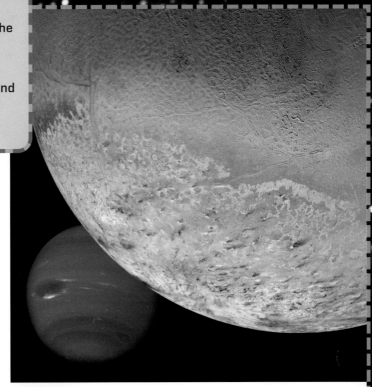

Neptune has one large **moon** and 10 small moons. The large moon, called Triton, was discovered in 1846.

Neptune has only been visited by one spacecraft, the **space probe** *Voyager 2*, in 1989. The space probe took photographs of Neptune and gathered information about it. The *Hubble Space Telescope* has also given us much useful information about Neptune.

Because of its blue color, the planet Neptune was named after Neptune, the Roman god of the sea.

◀ Neptune, god of the sea

The Eighth Planet

Neptune is part of the solar system, which consists mainly of the Sun and nine planets. The planets **revolve** around the Sun. Neptune is the eighth closest planet to the Sun.

The solar system also has comets and asteroids moving around in it. Comets are large balls of rock, ice, **gas**, and dust which orbit the Sun. Comets start their orbit far away from the Sun. They travel in close to the Sun, go around it, and then travel out again. When they come close to the Sun, comets grow a tail.

Asteroids are rocks. There are millions of asteroids in the solar system. They can be small or large. The largest asteroid, named Ceres, is about 584 miles (940 kilometers) across. Most asteroids orbit the Sun in a path called the asteroid belt, between the orbits of Mars and Jupiter.

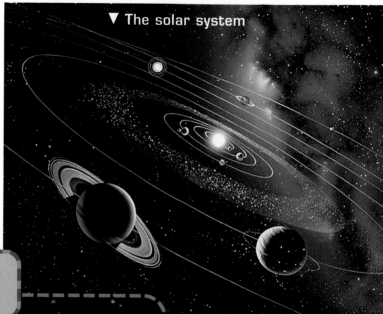

▼ The solar system

The solar system is about 4,600 million years old.

The planets in the solar system are made of rock, ice, gas, and liquid. Mercury, Venus, Earth, and Mars are made of rock. Pluto is probably made of rock and ice. These are the smallest planets.

Jupiter, Saturn, Uranus, and Neptune are made mainly of gas and liquid. They are the largest planets. They are often called the gas giants, because they have no solid ground to land on.

Planets, comets, and asteroids are lit up by light from the Sun. They do not make their own light the way stars do.

▶ The planets, from smallest to largest, are: Pluto, Mercury, Mars, Venus, Earth, Neptune, Uranus, Saturn, and Jupiter.

Planet	Average distance from Sun	
Mercury	35,960,000 miles	(57,910,000 kilometers)
Venus	67,190,000 miles	(108,200,000 kilometers)
Earth	92,900,000 miles	(149,600,000 kilometers)
Mars	141,550,000 miles	(227,940,000 kilometers)
Jupiter	483,340,000 miles	(778,330,000 kilometers)
Saturn	887,660,000 miles	(1,429,400,000 kilometers)
Uranus	1,782,880,000 miles	(2,870,990,000 kilometers)
Neptune	2,796,000,000 miles	(4,504,000,000 kilometers)
Pluto	3,672,300,000 miles	(5,913,520,000 kilometers)

The name "solar system" comes from the word "Solaris." This is the official name for the Sun. The Sun is a star.

7

On Neptune

As it travels around the Sun, the blue planet Neptune spins on its **axis**.

Rotation

Neptune **rotates** on its axis once every 16.11 Earth hours. This is a very fast spin for such a large planet. Earth is much smaller and takes 24 hours to rotate once. Neptune does not rotate in a perfectly upright position. Its axis is tilted over by 30 degrees.

Neptune gives off pulses of **radio signals**. Astronomers worked out how fast Neptune rotates by studying these radio signals.

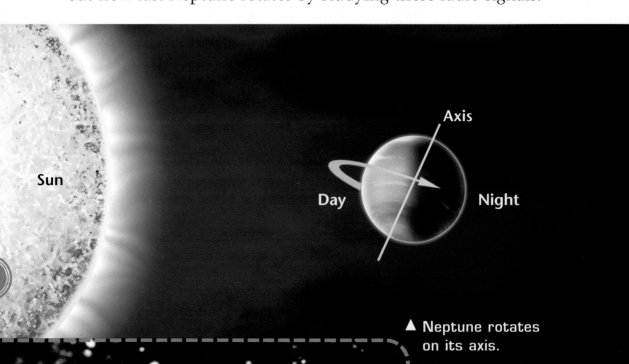

Sun

Day

Axis

Night

▲ Neptune rotates on its axis.

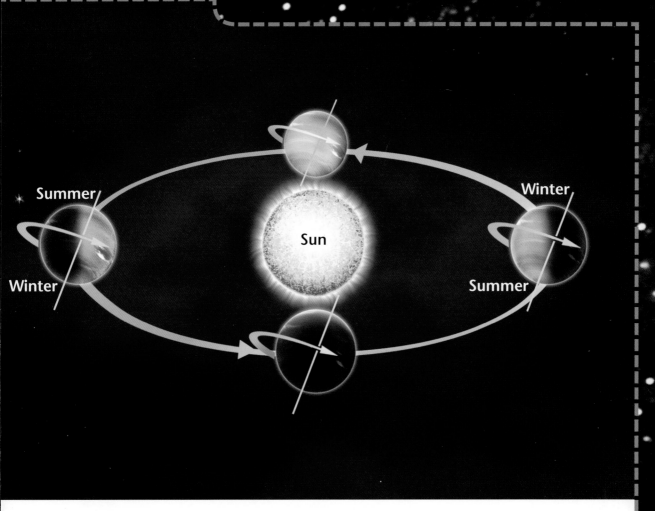

Summer

Winter

Sun

Winter

Summer

▲ Neptune's year, as it revolves around the Sun

Revolution

Neptune revolves around the Sun in an almost perfect circle. It takes 164.79 Earth years to orbit the Sun once, and this is the length of Neptune's year. The Sun's strong **gravity** keeps Neptune revolving around it.

Because its axis is tilted, Neptune has four seasons as it revolves around the Sun. Each season is about 41 years long. Astronomers have not found changes on Neptune that are caused by the seasons.

Size and Structure

Neptune is the fourth largest planet and the smallest of the four gas-giant planets. Its **diameter** at the **equator** is 30,759 miles (49,532 kilometers).

Neptune is heavier than the planet Uranus, making it the third heaviest gas-giant planet. Neptune contains a fairly small amount of the substance hydrogen, the lightest substance that there is. Neptune is made mainly of heavier substances, which is why it is heavy for its size.

▼ Compare the size of Neptune and Earth.

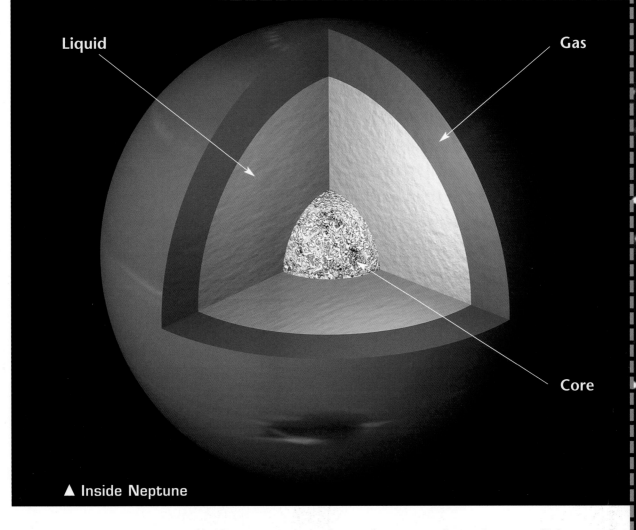

Liquid

Gas

Core

▲ Inside Neptune

Neptune is called a gas-giant planet because there is no solid ground to land on. Its outer third consists of gases. This layer is Neptune's **atmosphere**. The atmosphere is mostly hydrogen and helium, with a small amount of methane. The methane causes Neptune's blue color.

Neptune is liquid below the atmosphere. The atmosphere changes gradually into the liquid below it. The liquid contains mainly the substances ammonia, methane, and water. There is a **core** at the center of Neptune made of liquid rock. The core is about as heavy as Earth.

Atmosphere

Neptune's atmosphere is about 85 percent hydrogen, 13 percent helium, and 2 percent methane. The planet's blue color comes from the methane in the atmosphere. Neptune is such a bright blue that there must be something else in the atmosphere as well as methane making the blue color. Methane would normally make a greenish-blue color.

Neptune has bands of clouds in its atmosphere. Most of the cloud bands blow towards the west, the opposite direction to Neptune's rotation. The bands blow at different speeds, with the fastest winds blowing around the equator at up to 1,200 miles (2,000 kilometers) per hour. These are the fastest winds in the whole solar system.

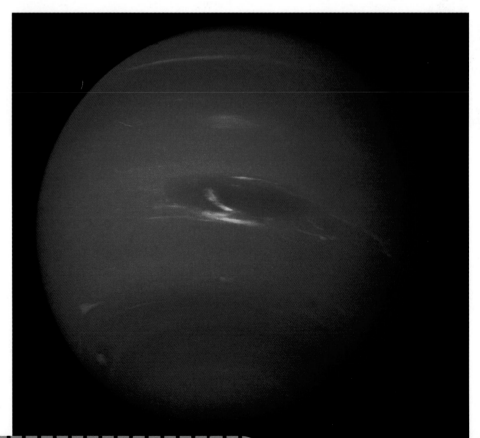

▶ Neptune has clouds.

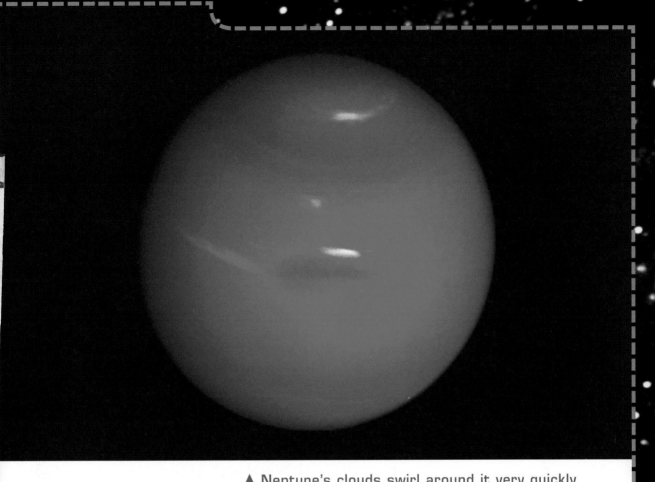

▲ Neptune's clouds swirl around it very quickly.

Some of Neptune's clouds are very long and thin. They form high up in the atmosphere and are probably made of **crystals** of methane ice. Lower down in the atmosphere is the main cloud layer, which is probably made of small drops of methane. Neptune has several large, dark spots which are similar to **hurricanes**. They come and go from time to time.

It is not known what drives Neptune's very fast winds and huge storms. On Earth, the weather is caused by the Sun heating up the air and the oceans. On Neptune, the Sun is 900 times dimmer than it is on Earth. There seems to be something else besides heat from the Sun that causes Neptune's weather.

13

Great Dark Spot

The space probe *Voyager 2* photographed Neptune in 1989. At that time, Neptune had a huge, dark-colored storm cloud which was named the Great Dark Spot. Pictures taken by the *Hubble Space Telescope* in 1994 showed that the Great Dark Spot had disappeared. Soon afterwards, a new dark spot was discovered.

The Great Dark Spot had feathery white clouds around its edge. These clouds were made from methane gas which had been pushed up high in the atmosphere. The gas then cooled and turned into methane ice crystals. The dark spot was possibly an area of clear gas which showed the cloud layer below.

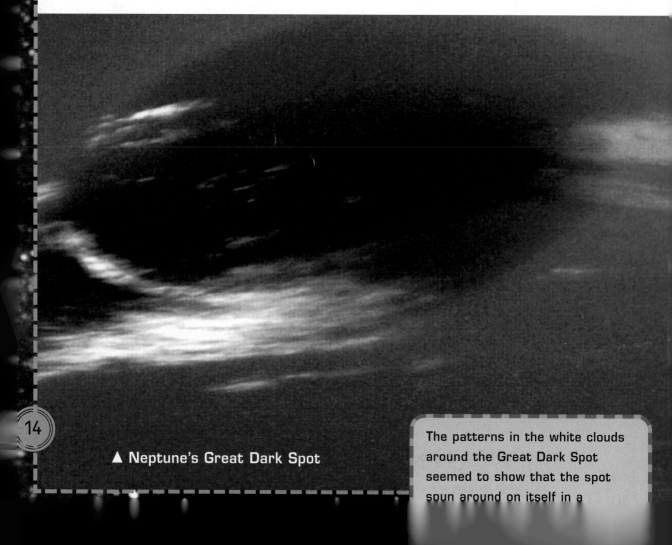

▲ Neptune's Great Dark Spot

The patterns in the white clouds around the Great Dark Spot seemed to show that the spot spun around on itself in a

◀ These streaks of cloud are being blown along by strong winds. They are making shadows on the blue atmosphere below.

▶ This picture, taken in 1989, shows the Great Dark Spot. Below it is a white cloud, nicknamed "The Scooter," which traveled around Neptune once every 16 hours. Underneath it is a small dark spot with a white center.

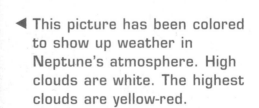

◀ This picture has been colored to show up weather in Neptune's atmosphere. High clouds are white. The highest clouds are yellow-red.

15

Giving Off Heat

Neptune gives off more than twice as much heat as it receives from the Sun. It must be making its own heat inside, as well as being heated by the Sun.

Astronomers are not sure how Neptune makes its heat. It is possibly caused by the gases in the atmosphere circling around. The gases rise up, move towards the equator or **poles** and cool down, then sink and heat up again. These movements would make gases squeeze together, which would make them heat up.

▼ Gases move around inside Neptune's atmosphere.

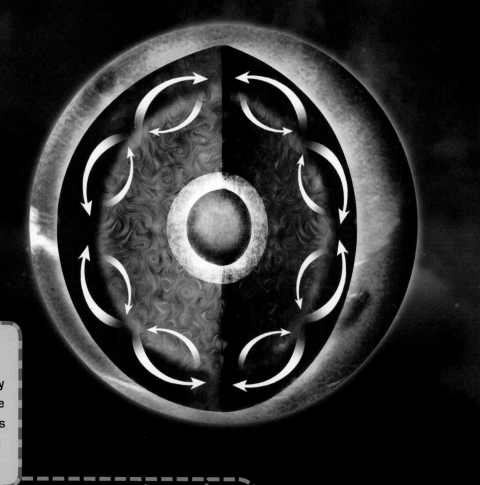

The heat that is made inside Neptune probably helps to drive the very strong winds that blow around the planet.

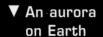
▼ An aurora
on Earth

Neptune has **auroras**, which are caused by the planet's magnetic field. The auroras happen over large parts of the planet, not just at the poles like on Earth.

Magnetic Field

Like Earth, Neptune has a magnetic field in parts of space around it. A magnetic field is an area which makes magnetic particles move. The magnetic field is possibly caused by liquids inside Neptune spinning as the planet spins.

Neptune's magnetic field was discovered by the space probe *Voyager 2*. It detected pulses of radio signals being given off by Neptune. The radio signals are caused by **charged** particles trapped in Neptune's spinning magnetic field. The radio signals were used by astronomers to work out that Neptune's day is 16 hours and 7 minutes long.

Rings and Moons

Neptune has at least 11 moons circling around it, as well as rings of dust.

Rings

Neptune has four main rings around it. The rings do not show up in ordinary photographs of Neptune because they are very thin and very dark.

Neptune's rings were first detected from Earth with telescopes, in the mid-1980s. The rings blocked out light from stars behind them, as Neptune passed in front of the stars. The rings appeared to be incomplete, and only went part of the way around Neptune. They were called arcs. The space probe *Voyager 2* went to Neptune in 1989. Its pictures showed that the rings went right around Neptune.

Neptune's rings are probably made from dust thrown up when asteroids have hit Neptune's moons.

▲ An artist's impression of Neptune, its rings, and the moon Triton

▼ Adams ring

The Adams ring was named after the astronomer Adams. He helped discover Neptune, working out where Neptune could be found when he was 24 years old.

The ring closest to Neptune is a cloudy ring called Galle. Galle reaches down to the cloud tops of Neptune. Outside Galle is the ring named Leverrier. Leverrier is a brighter ring in pictures. Leverrier has a faint extension outwards called Lassel. Lassel has a faint ring on its outside edge called Arago.

The outside ring, about 39,080 miles (62,930 kilometers) out from Neptune, is called Adams. It is a brighter ring. Adams contains five arcs called Liberty, Equality 1, Equality 2, Fraternity, and Courage. These arcs are brighter sections of the Adams ring. Part of Adams has a twisted appearance.

Moons

Neptune has 11 moons. One moon, named Triton, is large. Two moons, Proteus and Nereid, are small. The other eight moons are tiny. Triton and Nereid can be seen from Earth with telescopes.

Proteus and five tiny inner moons were discovered by the space probe *Voyager 2* in 1989. Very little is known about Neptune's five tiny inner moons. They have **irregular** shapes, are dark in color, and range from about 40 to 120 miles (60 to 200 kilometers) across. Four of these tiny moons orbit Neptune amongst the rings. One tiny moon orbits Neptune outside the rings, but closer to Neptune than the three larger moons.

Three tiny moons outside the others were discovered in 2002. Little is known about these moons.

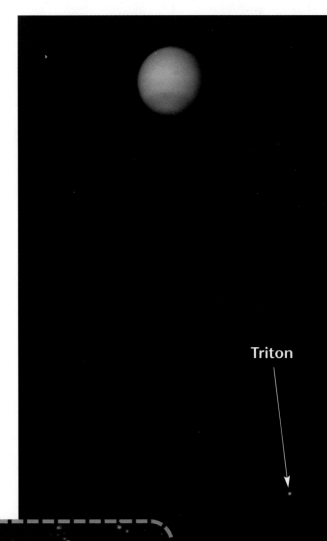

Triton

▶ Neptune and Triton

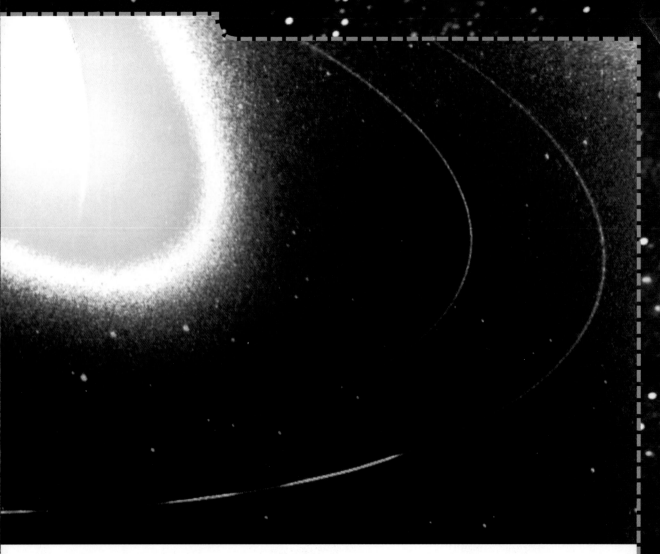

▲ The shepherd moons in the rings of Neptune

Shepherd Moons

Two of Neptune's tiny moons seem to be shepherd moons. Shepherd moons orbit a planet close to a ring and help to keep the ring in place. They stop particles from flying out too far, or from falling in towards the planet.

The tiny moon Galatea orbits Neptune just inside the Adams ring. It appears to be a shepherd moon for Adams. The moon Despina seems to be a shepherd moon for the ring Leverrier. Despina orbits just inside the Leverrier ring.

Triton

Triton is Neptune's only large moon. It is 1,700 miles (2,700 kilometers) in diameter. Triton was discovered in 1846, only a few weeks after Neptune was discovered.

Triton orbits Neptune in 5.88 Earth days, in the opposite direction to the direction in which Neptune spins. Only a few other moons in the solar system do this. It shows that Triton was probably once floating freely in space and was pulled in to Neptune by the planet's strong gravity.

Triton's orbit is 220,000 miles (355,000 kilometers) from Neptune, but it is slowly spiraling in towards Neptune and will probably one day break up and make rings as spectacular as Saturn's rings.

▼ Triton

▶ Triton was a Greek god of the sea. He was the son of Poseidon (the Greek name for Neptune) and is usually shown with a man's body and the tail of a fish.

The temperature on Triton is a very cold –390 degrees Fahrenheit (–235 degrees Celsius). This low temperature makes Triton the coldest known object in the solar system. Triton is even colder than the planet Pluto, which is much further away from the Sun.

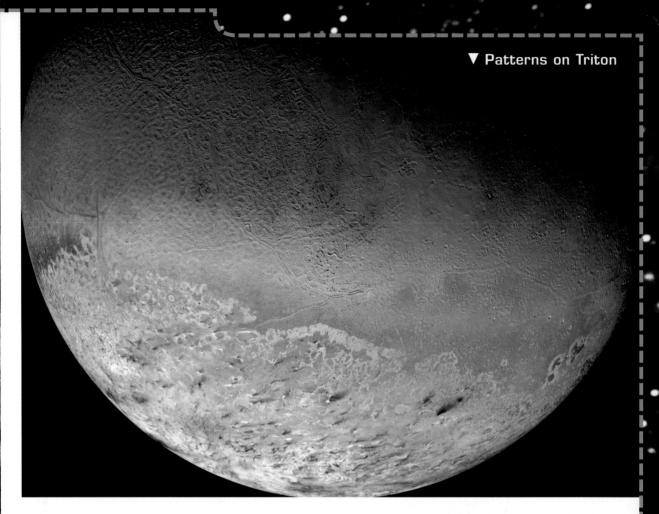

Triton is about 25 percent frozen water and 75 percent rock. It has a very thin atmosphere of mostly nitrogen, with some methane. Triton has a similar make-up to the planet Pluto.

Triton has very few **craters**, which shows that Triton's surface is constantly being renewed. Large parts of Triton have ridges and valleys making patterns on its surface. These parts are called "cantaloupe terrain," because they look something like a cantaloupe melon.

Triton is partly covered with pink ice. It has active **volcanoes** which throw dark dust up to 5 miles (8 kilometers) high. This blows across the surface of Triton, leaving dark streaks on the pink ice.

Proteus

Proteus is a small moon measuring about 260 miles (418 kilometers) across. It is Neptune's second largest moon. Proteus is not shaped like a ball, but has an irregular shape. Its orbit is about 73,000 miles (118,000 kilometers) from Neptune.

Proteus is as black as soot. It is one of the darkest objects in the solar system. Proteus is larger than Neptune's moon Nereid, but was discovered after Nereid. This is because Proteus is dark and close to Neptune, and was lost in Neptune's glare. Proteus appears to have craters and grooves on its surface.

Triton orbits Neptune in 5.88 Earth days.
Proteus orbits Neptune in 26.9 Earth hours.
Nereid orbits Neptune in 360 Earth days.

▼ Proteus

▶ Proteus was a sea god who could change his shape when he wanted. The human shown here is trying to grab Proteus in order to learn the secret of the future.

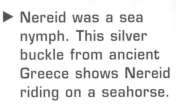

▶ Nereid was a sea nymph. This silver buckle from ancient Greece shows Nereid riding on a seahorse.

Nereid

Very little is known about Nereid, which orbits 3,421,000 miles (5,509,000 kilometers) from the planet. It was discovered in 1949. Nereid is the third largest of Neptune's moons, measuring about 210 miles (340 kilometers) across.

Nereid takes 360 Earth days to orbit Neptune. Its orbit has a very strange shape. Sometimes Nereid comes closer to Neptune, and sometimes it is much further away. Nereid has a stranger orbit than any planet or moon in the solar system. The strange orbit shows that Nereid was probably once floating freely in space, and was pulled towards Neptune by the planet's gravity.

25

Exploring Neptune

Neptune has been visited by one spacecraft without people on board, called *Voyager 2*. The spacecraft was operated by astronomers on Earth by sending and receiving radio signals. This type of spacecraft is called a space probe.

Voyager 2 was launched in 1977. It visited Jupiter in 1979, Saturn in 1981, Uranus in 1986, and Neptune in 1989. The probe is on its way out of the solar system. Astronomers will keep getting information from it until about 2030.

Voyager 2 uses its dish-shaped antennas to send radio signals back to Earth.

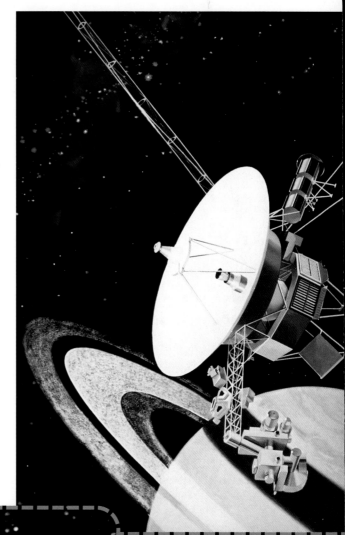

▶ *Voyager 2*

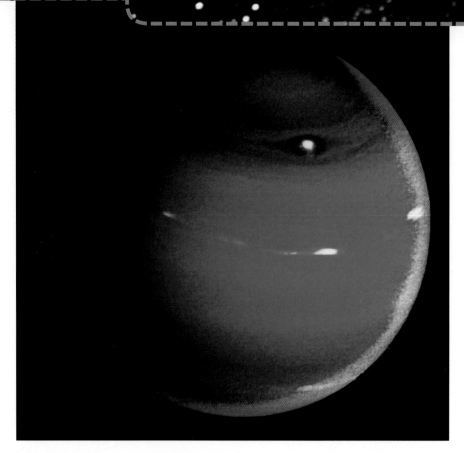

▲ This false color picture of Neptune was taken by *Voyager 2*. The red color shows the methane in Neptune's atmosphere.

Voyager 2 Discoveries

In 1989, *Voyager 2* discovered six of Neptune's moons and found that Neptune's rings went right around the planet. The space probe photographed clouds on Neptune, including the Great Dark Spot. It also found that most of Neptune's winds blew towards the west.

Voyager 2 measured heat coming from Neptune and found that the atmosphere was warmer at the equator and south pole, and cooler in between. It discovered pulses of radio waves coming from Neptune. *Voyager 2* also discovered the unusual surface of the moon Triton, and the volcanoes on Triton.

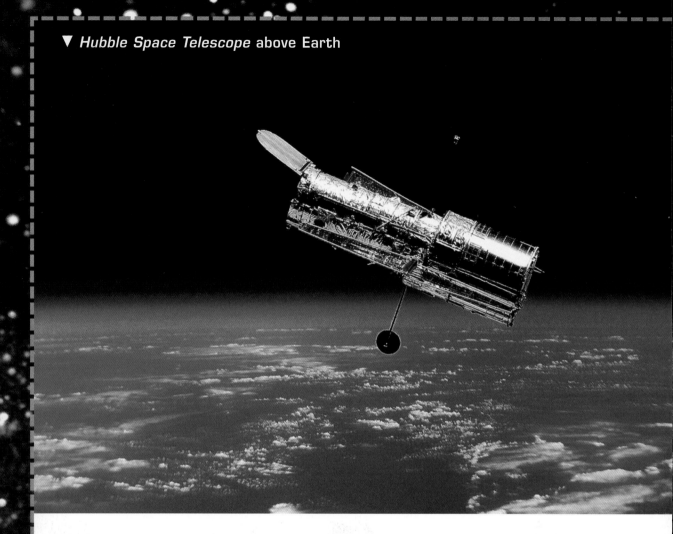

Hubble Space Telescope

The *Hubble Space Telescope* (*HST*) is a telescope which orbits Earth. It gets a clearer view of stars and planets than telescopes on Earth because it is above Earth's atmosphere. The *HST* carries cameras and instruments for detecting heat and studying light. The *HST* was sent into space on board the space shuttle *Discovery* in 1990. When *Discovery* reached space, the *HST* was released to orbit Earth on its own.

The *HST* has taken very clear pictures of Neptune's clouds. These pictures show that the storms and clouds on Neptune change often.

Questions about Neptune

There is still a lot to learn about Neptune. One day, astronomers hope to find out the answers to questions such as these:

- Why does Neptune not contain very much hydrogen and helium?
- Why does Neptune have such strong winds? Where does the energy to drive these winds come from?
- Why do we not see the Great Dark Spot today? Has the storm died down, or is it covered up by the atmosphere?
- Why does Nereid have such an unusual orbit?
- What causes the volcanoes on Triton?
- Was Triton formed in the same way as the planet Pluto?

▶ An artist's impression of a geyser on Triton, with Neptune in the background

Neptune Fact Summary

Distance from Sun (average)	2,796,000,000 miles (4,504,000,000 kilometers)
Diameter (at equator)	30,759 miles (49,532 kilometers)
Mass	17.15 times Earth's mass
Density	1.76 times the density of water
Gravity	1.12 times Earth's gravity
Temperature (at cloud tops)	–364 degrees Fahrenheit (–220 degrees Celsius)
Rotation on axis	16.11 Earth hours
Revolution	164.79 Earth years
Number of moons	11 plus

Web Sites

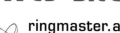

ringmaster.arc.nasa.gov/neptune/neptune.html
Neptune's ring system

www.nineplanets.org/
The nine planets—a tour of the solar system

www.enchantedlearning.com
Enchanted Learning web site—click on "Astronomy"

stardate.org
Stargazing with the University of Texas McDonald Observatory

pds.jpl.nasa.gov/planets/welcome.htm
Images from NASA's planetary exploration program

Glossary

ancient from thousands of years ago

astronomers people who study stars, planets, and other bodies in space

atmosphere a layer of gas around a large body in space

auroras curtains or bands of light in the sky

axis an imaginary line through the middle of an object, from top to bottom

charged carrying electric energy

core the inside, or middle part of a planet

craters bowl-shaped holes in the ground

crystals tiny pieces of pure substance

density a measure of how heavy something is for its size

diameter the distance across

equator an imaginary line around the middle of a globe

gas a substance in which the particles are far apart, so they are not solid or liquid

gravity a force which pulls one body towards another body

hurricanes storms with violent wind, blowing in a spiral

irregular not evenly shaped, not globe shaped

mass a measure of how much substance is in something

moon a natural body which circles around a planet

orbit *noun* the path a body takes when it moves around another body *verb* to travel on a path around another body in space

planet a large body which circles the Sun

poles the top and bottom of a globe

radio signals invisible rays

revolve travel around another body

rotates spins

space probe a spacecraft which does not carry people

stars huge balls of glowing gas in space

telescope an instrument for making far away objects look bigger and more detailed

volcanoes holes in the ground through which lava flows

31

Index

32